To Clay -

Stay Curious Phewburt

Clayton

This page is to signify my best friend. In 2019, I along with many others lost someone close to us; his name was Clayton and he was my best friend. We spent nearly everyday together and I can't thank him enough for the joy he brought me and everyone around him. To those who knew him, I'm sorry. It's something that would've been difficult to stop, but if I could've I wouldn't have hesitated. Remember who he was and the joy brought us all. This is a tribute to him and his name. Make sure that if you're going through a rough time that you surround yourself with good people, get aid of some kind, or reach out to others—it will get better; tell everyone you love that you love them. Make sure they know it, and be kind, we're all human.

Stay Curious Clayton

CONTENTS

Introduction

The Universe we live in is vast; open to all to view and wonder upon. There is nothing more beautiful than pondering on what exists among the stars and the cosmos and how our lives relate to that of the Universe and divine creation. Being curious is what makes us, us and allows us to enjoy life with only our minds. There is so much to explore and so much to discover that humans have yet to even encounter. Our minds, however, allow us to fathom what might be new discoveries before they even exist in our realm. Thus the curious hold the ultimate power of wonder and creativity. This is a super power all can enjoy, and one we must all have. Curious fly beyond the stars and hover among the cosmological beings of exploration. We as the curious seek knowledge and find joy in understanding anything and everything the world and its relatives has to offer. Don't stop wondering, it is a joy to life and a joy for us.

Preface

Feel free to skip this page and continue on to the wonderful world of curiosity but I thought it would be right to tell you all a little about myself before we embark on this incredible journey.

My name is Logan and I, at the moment of writing this, am an 18-year-old undergraduate student who has dreamt of writing a book for years. This is not my first attempt at writing a book, the others have been scrapped and may become books later on. I enjoy learning a bit too much and I hope to share that trait with some of you in this book. Do enjoy the pages and feel free to skip around as you wish, this book was intended to have no order and to allow for a free choice at whichever adventure you wish to take. And always remember:

Stay Curious

-        Logan Corrente

Chapter 1

Solar system

Earth

W e start off our journey at the place it all began—for us at least. One of the most beautiful things ever created, our own planet is astonishing in the sights it contains. Even though Earth is undeniably massive, it is nothing relative to the size of Jupiter or even the Sun. But the beauty it holds is truly astonishing. To think that all life as we know it began on this very rock is astounding. Mother Nature truly has her ways of creating such a wonderful place. How many more Earth-like planets are there? What secrets do they hold? So many undiscovered mysteries just waiting to be found; yearning to be discovered across the Universe.

The Moon

Our own natural satellite*, the Moon, or Luna†, is the first large ball of matter we visited as a species other than Earth. The Moon's orbit is about 27 days, meaning it takes this long for it to orbit around Earth. However, it also takes 27 days to rotate completely on its axis, which means for us average observers, the Moon doesn't seem to be rotating at all. It is interesting to see the moon in its entirety—floating seamlessly in the night sky and even during the day as if it were planted in the sky by another being. It truly is a beautiful sight and I would argue that the moonmen and moonwomen would enjoy a vacation to Earth for a bit while we ventured on their lunar surface, but I guess we'll see what they think.

* A celestial body orbiting another of larger size.

† Defined in Latin as: "silver as used in alchemy."

The Sun

O n our journey through the solar system I figured we would take a trip to the place that allows life to exist: The Sun. Without the sun, the sky would always be dark, photosynthesis* would not be able to occur in plants, and animal life would not be able to survive. This wonderful ball of glowing plasma is incredibly important to all life on Earth.

It's incredible to think that this massive sphere is home to billions of nuclear explosions every second and this is the reason we have sunrises and sunsets; why we have plants and why we get to exist on this wonderful planet. Perhaps the atomic physicists of the 20th century took this potency a bit too lightly.

---

* Synthesis of chemical compounds with the aid of radiant energy and especially light.

† A collection of charged particles (as in the atmospheres of stars or in a metal) containing about equal numbers of positive ions and electrons and exhibiting some properties of a gas but differing from a gas in being a good conductor of electricity and in being affected by a magnetic field.

Mercury

The closest planet to the Sun in our solar system. Mercury is very tiny and very hot due to its proximity relative to the Sun. It is not, however, the hottest planet in our Solar System, that award goes to Venus which is roughly 50° (celsius) more than Mercury.

In billions of years Mercury will be entirely submerged into the Sun once it becomes a Red Giant*. While this won't happen for quite some time it is still interesting to consider the state of the Universe at such a time: Mercury is on the verge of death and the Earth is likely completely dry. Who knows what the humans will be up to; probably debating religion.

---

* A star that has low surface temperature and a diameter that is large relative to the sun.

Venus

The second planet closest to the Sun, Venus is practically invisible at its surface due to the dense clouds at the top of its atmosphere. Unlike the other planets in our solar system, which travel in a counterclockwise rotation, Venus travels in a clockwise rotation. I would advise not to take any trips to Venus, as you would most likely burn to a crisp upon entering its atmosphere. However I do think it would make for a great story to be the first person to have video footage of the atmosphere of Venus. As sad as it is to say, the surface probably looks like some of our oceans.

Earth II

E arth is definitely one of my favorite planets; disregarding the many pestering people that inhabit our planet, we live on a sphere of great wonder. Imagine a world without people, without animals or the most intricate or simplistic organisms. It would be a world just like Mars; with only dust and matter. But, if we add just the right ingredients then we get what we have: life. How could this have come to be? On Earth, at least, a question that has challenged the minds of the curious for millennia. Many conclusions can be derived from such a question. Any suggestion may be considered but none factual for certain as the only one who witnessed the birth of life was the one who created life itself. I have no place in defining such a time, but I can express curiosity at such a thought.

Our planet is something special; whether you see it in that light or not, it is remarkable that we exist on this planet. The different possibilities of life, solar system size, planet size, planet color, different star, yet all of our achieved possibilities seem to be in perfect conditions to host us. Perhaps one day we will move from Earth onto Europa or Titan*. Hopefully; I would like my own hut on Europa.

---

* Europa is a moon that orbits Jupiter and is one of the most well known moons relative to its neighbors. Titan is a famous moon of Saturn that falls under the same category for the popular planet.

Mars

Do you believe there was life on Mars? I'd like to say yes; with the somewhat sustainable atmosphere as well as ice, I don't see how there couldn't be. I can only imagine what they would look like or how they would act; if there were any it doesn't seem that they were intelligent or at least if they were they did not evolve all that much.

It would be a dream to travel to Mars one day; sounds like it would be nice and peaceful and I could eat my soup in peace. Who knows, maybe Elon* will give me a ride to the Red Planet.

To think of the possibility of life on another planet other than Earth is dumbfounding as it has never been seen before, but I do hope we see it soon. I would love to greet a martian.

* A wonderful, professional nerd and engineer.

Jupiter

Moving away from the inner-planets and on to the outer-planets, we arrive at Jupiter. The largest planet in our solar system; Jupiter, is basically a massive shield for the earth: blocking major asteroids and space debris that could cause serious damage to our home. Most of you probably already know about "the spot", but it is a giant storm that has been raging on the planet for millennia. Jupiter is basically the layer-cake of our solar system.

A beautiful sight to see, however frightening as the Universe has the ability to create things of such gargantuan size, and Jupiter does not even come close to some of the largest out there.

Saturn

Saturn is iconic for its rings, although it is not the only planet that has rings, Saturn's rings consist of asteroids and space debris that are caught in its gravitational pull.

It is interesting to see how sci-fi portrays Saturn as it almost seems to typically be a rest-stop of our Solar System with fuel stations and restaurants existing and thriving among the asteroid belt. I must say I would love to find there to be a wonderful place that sells intergalactic chicken-fingers not but a few hundred thousand light-years from Saturn's surface.

Uranus

U ranus has given us many jokes over the years due to it's wonderful name. Unknown to many, Uranus actually has rings; they just aren't as popular as Saturn's. Furthermore Uranus has one of the most odd rotations since it rotates as if it is rolling on the ground rather than spinning. A strange planet no doubt, but still as incredible as the rest.

It would be nice to see planets like Uranus and Neptune be the stars of movies and TV shows for a change; while they are entirely gas I do think it could make for a great story if someone were to get lost among the atmosphere.

Neptune

N eptune is the second farthest planet from the sun, for all you Pluto fans. This big blue ball of gas looks quite like a giant ocean, hence the name: Neptune. Much like the other gaseous planets, I would not recommend trying to visit this one. You might fall through.

Neptune is a beautiful planet and is a favorite among many of my acquaintances. I know it is quite far away, both literally and chronologically, but I hope sometime soon we get to see more exploration of the outer planets; there is still so much we have yet to learn about these incredible balls of gas that lay next to us in our Solar System.

Pluto

L ast but certainly not least, we have Pluto*. Unfortunately, Pluto is not necessarily a planet, but a dwarf planet rather considering its small size relative to the other planets. Pluto is practically a ball of ice and other isolated elements. Pluto is the coldest planet in our solar system with a temperature of a cool -223° celsius.

I personally would give Pluto the title of 'Planet' even though it goes against practically everything astrophysicists and astronomers have argued about for decades. Maybe soon we will send a probe to Pluto to let it feel at least a bit of recognition.

---

* A planet.

Chapter 2

Wonders of the Universe

The Observable Universe

W e are within the Observable Universe, quite beauti-ful to be honest. Since the Universe is theorized to be ~14.6 billion years old, the distance across the entire Universe would be 14.6 billion light-years across. Of course this is theoretical since we can't yet prove that the Universe is spherical, or that it even has an end or if there is something beyond that end. Think of this, the average person is a little over 5 feet tall (slightly over 1.5 meters), the Universe is about 9.223×10^{18} feet in diameter (about 1.38×10^{27} meters) (given the theory is true) this is how microscopic we really are.

Black Holes

B lack Holes have been theorized to be massive holes that allow nothing to escape, swallowing everything in its path, even light. It has also been theorized that black holes decay slowly over time. This period of time is incredibly long and nothing will exist except for the black holes at this point, but there are particles that are said to pop in and out of existence all over the Universe. Some of these particles pop into existence at the Black Hole's event horizon, upon forming they collide with another particle, disappear, and the cycle continues*. This action causes deterioration at the event horizon, it's just incredibly slow. It's like watching grass grow, if grass grew an inch every 10 billion years.

* Also known as Hawking Radiation.

Black Holes II

T he first ever image of a black hole was released not too long ago, and it almost appears to be devouring a star of some sort. Nonetheless it amazes me how technology has allowed us to capture an image of a body of mass that not even light can escape, which would bring no reflection of the black hole making it seem invisible; yet here is a picture of it. This also proves yet again Einstein's theory of general relativity in which he had theorized the existence of black holes as well as white holes, although white wholes still may be hiding waiting for their photoshoot.

Wormholes

For everyone that enjoys sci-fi movies, this one should be pretty self-explanatory. A wormhole is essentially a connection between two points on the space time continuum that, theoretically, can be entered and you will arrive on the other side. This, of course, has not been proven yet and the only thing we know of that can warp space time is mass. This would be somewhat of a doorway to a different part of the Universe. Given that we do discover a wormhole, where will it take us? It's not like we can use GPS or look it up online. Perhaps there is a mathematical proof to this that remains undiscovered.

Galaxies

F or all of you Nintendo fans, I hope you've played Super Mario Galaxy*, for no reason really I just love the game. Besides that, galaxies are incredible; massive black holes live at the center, sometimes even two, with gravitational forces beyond comprehension, they hold all of their billions of stars, planets, etc. in one place millions of light years across. The amount of planets and bits of cosmic matter is incomprehensible. And it is to that I say how can there not be other forms of life out there? We cannot even fathom the Universe we live in perhaps there are others who can; I don't feel as though we are the only ones.

* A popular game released by Nintendo in 2007.

Pulsars

I f there were a giant radio in space, this would be it. Pulsars are Neutron Stars that spin incredibly fast that emit beams of electromagnetic radiation*. This emission occurs because the Neutron Stars are spinning so fast; their spin causes the generation of a very strong magnetic field. Charged particles accelerate along the magnetic field and are launched into space at high velocities.

If we needed a source of energy that wasn't a star, I suppose we could go to a pulsar. The energy released during each pulse could be absorbed by something similar to a Dyson Sphere or anything surrounding the celestial object.

---

* A series of electromagnetic waves†

† One of the waves that are propagated by simultaneous periodic variations of electric and magnetic field intensity and that include radio waves, infrared, visible light, ultraviolet, X-rays, and gamma rays.

Storm of Jupiter

L et's take a break from arbitrary philosophy and venture back into the world of arbitrary facts and knowledge.

Let's come back and look at the infamous storm that has been raging upon its gaseous surface for 150+ years. This storm is described as an anticyclone* and has reached speeds of 300 mph (483 kph). To put that into perspective: the most devastating hurricane on Earth was hurricane Wilma; Wilma reached speeds of 183 mph (295 kph) and tore apart houses, flipped cars and trucks, decimated trees and power lines, etc. . With that being said, maybe don't let little Timmy fly his new batman kite on your trip to Jupiter.

We can only hope that on Earth mother nature doesn't just throw a 300 mph hurricane at us any time in the future. But considering Jupiter has no solid matter, and is only a giant ocean of gas and more gas, then, well, it makes sense why there is an extremely large area of gas that does not react very well together; thus making a very pretty, yet very dangerous, red dot.

---

* A system of winds that rotates about a center of high atmospheric pressure clockwise in the northern hemisphere and counterclockwise in the southern, that usually advances at 20 to 30 miles (about 30 to 50 kilometers) per hour, and that usually has a diameter of 1500 to 2500 miles (2400 to 4000 kilometers).

Celestial* Beings

While humans have yet to prove their existence, it should not be ruled out that the Universe may have been created by a being more potent than we could ever imagine. To see the perfection of how the Universe is built is so incredibly fascinating and it makes you wonder how such a thing could be created by just chance. There must've been some other force or some other magnificent being that sits among the Cosmos and guides the Universe into perfection. Or perhaps a turtle does all of this on Earth. To know would be to unlock the answer to the Universe and I feel that is something humans are definitely not yet ready for.

* Of, relating to, or suggesting heaven or divinity.

Cosmos

T he Universe itself is glorious. As we stare up at the Cosmos we see curiosity, glimmering lights, and gorgeous patterns that have caused scientists to study them for centuries.

The most incredible thing in the Universe, is the Universe itself. There is nothing more complex yet so beautifully simple anywhere that we would be able to comprehend. So in our minds, or in my mind I suppose, the Universe is everything; it created and destroyed all that has ever been and will be. We live among an entity that has all of this power, yet it seems as though it is just a black box filled with little balls of light or dark. Perhaps a toddler lost his marbles, literally.

Nebulae

N ebulae are essentially massive clouds of matter that are collected via gravity after things like supernovae. These dust clouds are one of the most beautiful sights in the Universe; each one taking a different shape and consisting of different elements and colors.

After millions of years, however, these works of art come together to form planets, solar systems, and even galaxies. They can be seen as the recycle zones of the Universe; while we have Black Holes as the trash cans, the Nebulae take the unwanted or overused matter and create it into something magnificent again.

Chapter 3

Elements of the Universe

Light

L ight is incredible and the entire reason we can see any-thing at all is because of that strange little wave particle thing we call a photon*. An example of the incredible abilities of the Universe and light would be: When light passes behind a black hole, it warps. The light we see performing this action is the light that doesn't disappear into the black hole, instead it bends around the blackhole giving it the warped effect. If there was a Star directly behind a black hole, you would be able to see that star, this is because the black hole takes that light and pushes to its outer boundaries.

---

* A quantum† of electromagnetic radiation.

† Quanitity, amount.

Light II

L ight, the fastest entity known to man. Light travels at 186,000 miles per second. To put this in perspective, your car travels 60 miles per hour, given you're traveling at that speed.

Light is a particle that acts like a wave which makes it a unique part of physics, this particle is called a photon. The different frequencies of these light waves determines the colors we perceive. Going back to black holes, light cannot escape a black hole even though it travels faster than black-Friday shoppers at Best Buy. Which would mean black isn't necessarily a color; it's just nothing, but we give it a name so we at least have some sort of answer.

Time

T ime itself is an interesting concept. Considering Time in a logical view, it is just a form of measurement. There is no proof of a time space continuum or anything of the sort. [A time space continuum is essentially a timeline for the Universe, it houses the beginning middle and, well, the indefinite end of our Universe.] Time, as it relates to the Universe, is the 4th dimension. To be brief there are 10 different dimensions proposed in String Theory and time is the 4th dimension that coexists with space. However, is Time just a measurement? Or is it a relative distance that we can travel within and have a cup of tea with a velociraptor?

Time II

T ime: "the indefinite continued progress of existence and events in the past, present, and future regarded as a whole."

The undefined nature of time is intriguing yet terrifying. We live each day not knowing what our universe runs on. What fuels that which is the entity we live in? Perhaps time is the fuel. Maybe everyday we live is more fuel for the universe. But I feel if time were the fuel, then we are spending it. We are using the fuel to our advantage by existing, by taking up this time we may not have enough for everyone, or the Universe may have to find more. But if that is the case then time is precious, use the time you have and enjoy it. Spend it with care and choose to do things that are worthwhile. Spend time with family, and with friends. Pursue dreams and learn from experiences. To live is to spend time, and like everything, time is indefinite.

Matter

Matter is defined as "a physical substance in general, as distinct from mind and spirit..." Matter is everything; everything you touch, everything you see, and everything that exists. Well, apart from ideas, I guess. Matter can take many shapes; any shape actually given its a gas or liquid. Other than that matter can also take forms as a solid or a plasma. A solid being, a solid, just very a compact molecular structure that makes it difficult to alter the bonds. While a plasma is a state of matter in which both ions and electrons can coexist, also known as an ionized gas; some examples are lightning and the Sun, and modern theories state that this could be the most common state of matter in the universe.

Liquids can be thought of as sloppy solids, they have little structure but the bonds can be easily broken. With gases there is practically no structure and it is a jumble of particles that cannot be seen. Matter is the one thing in this universe that everybody on Earth knows of; the car you drive every day is a great combination of matter, the food you eat, the water you drink, the air you breathe: all matter.

Energy

What is energy? By definition, energy takes many forms. It can be the result of chemical reactions to fuel an organism, it can be the vitality required for physical or mental activity, or it can be a property of matter that allows things like movement or heat. Energy is given many names but what really is: energy? Einstein proposed the equation $e=mc^2$ (energy =mass x speed of light[2)], which has a longer form but simplified for this purpose, in which he stated that mass and energy are interchangeable; in other words, mass can be converted to energy, and energy can be converted to mass. How he discovered such a concept will forever be a confounding variable in my world of physics.

However the concept is quite intriguing. We've given energy so many different titles, but what exactly is it? Is energy just a form of matter that provides animation to our universe? And if energy cannot be created nor destroyed, then where did that first bit of energy that ignited it all come from? It seems we need another Einstein... or two.

Dark Energy

There isn't really much to say about this because, as of right now, we have no idea what exactly it does. However, theoretically, dark energy is everywhere; it's in your nose and everywhere else that you may be thinking. We just can't figure out it's purpose or what exactly "it" is, we just know that it's there. Physicists theorize that Dark Energy is the force of gravity, and if this were true then it may be the final piece to the puzzle that is The Theory of Everything. While we still have a multitude of decades to go before a discovery is made relating to this type of energy, it is still interesting to think about its power and what exactly it is capable of in the real world.

Entropy

R andomness: the quality or state of lacking a pattern or principle of organization; unpredictability.

The Universe is random. Everything can be predicted to a certain degree but no event has a certainty of 100%. Take, for instance, the electron of an atom; its position is random just as its velocity shares the same title. By the Uncertainty Principle, we may only know one for certain or the other, and the more we know of one the less we know of the other. This is: Entropy. Entropy is randomness; it is the uncertainty that is the universe. It is the not knowing of what will happen or how it will happen since it is unpredictable; it is uncertain. But just like everything else in the universe it has an opposite: negentropy.

Negentropy, instead of being disorder, is order, the lack of randomness in a system. To put it into famous words: "Perfectly balanced, as all things should be" - Thanos. Despite his intentions with the quote, there is truth to what he says. Deleting half of all living creatures definitely disrupted the balance of the universe, but it all came back to perfect balance in the end. Think of it like one of those old clown toys with the weight on the bottom, if you push it (causing disorder or increasing entropy) it will lean back and forth but eventually it returns back to its balanced state (causing order or increasing negentropy). Everything has its balance and it has its chaos, but it also has its order. Nothing would work without it.

Water

The essential, most important part of our lives is something so simple. A large bundle of H_2O molecules floating through the air and flowing down rivers and streams. But what makes this molecule so important? Why was it chosen to be the one that catalyzed life on Earth?

Scientists believe that all life began in the great oceans that conquered the Earth millions of years ago. This assumption is hard to overlook since there is such a plethora of beings in the ocean, some that we have yet to even discover. But what if we found these same oceans on another planet? Would we be able to find Nemo*? Hopefully there are many more oceanic planets in the Universe, the probability is just too high.

Next time you drink a cup of water just sit and enjoy the amount of living things you're drinking.

* A lovely clownfish from Disney.

Orbit

I t can be assumed that most pieces of space debris, satellites, any planet or star is locked within some sort of orbit. Due to the curvature of space-time, general relativity, planets of a higher mass give off a higher gravitational pull. Essentially the more mass within the entity, the stronger its pull. The Earth has an incredible amount of mass, and it curves space-time a significant amount; this results in other smaller entities getting caught in that curvature and end up orbiting around that larger body. Just like the Moon and the many satellites stuck in Earth's orbit. Many of these satellites will likely come back down onto Earth's surface due to the Earth's pull being stronger than the opposing pull of the satellite. Bodies like black holes have an extremely, incomprehensible amount of mass; which leads to the formation of a galaxy where trillions and trillions of bodies get stuck in an orbit around the center black hole of a galaxy because of its immense space-time curvature. Imagine placing a sheet of fabric over the opening of a large bucket, place a ball on top of the fabric, then rolling a smaller ball along the fabric and watch as the curvature created by the larger ball affects how the smaller ball rolls; it will cause the smaller ball to fall into an elliptical orbit and eventually land still next to the larger ball.

Chapter 4

Theoretical Concepts

Hawking Radiation

Black holes are extremely dense balls of nothingness that allow no matter to escape. Stephen Hawking, one of my favorite theoretical physicists, proposed an idea that at the event horizon black holes would catch a particle-antiparticle pair that would collide with one-another. This reaction causes the black hole to release radiation (Hawking Radiation). Which allowed scientists to conclude that the most potent, known, things in the Universe are dying indefinitely.

To think of how the Universe will be when the final black hole passes, and what will happen to all of its components after? Your guess is as good as mine.

Dyson Sphere

P roposed by Freeman Dyson, this massive structure is essentially an incredibly large solar panel. This structure would be capable of harnessing about 334 trillion Terawatts* of energy every second, considering that the Sun emits ~350 trillion Terawatts of energy per second and abiding by the laws of thermodynamics. To pit that into perspective: In 2014, the entire planet used 155,000 Terawatts in the whole year. This sphere would be capable of powering multiple planets in their entirety for hundreds of years.

This would make us a Type 2 civilization on the Kardashev scale† and would mean that we would have plenty of energy to stream netflix in 16k quality.

---

* A unit of power equal to one trillion watts.

† A method of measuring a civilization's level of technological advancement based on the amount of energy they are able to use. The measure was proposed by Soviet astronomer Nikolai Kardashev in 1964.

Multiverse Theory

Proposed by Stephen Hawking; Multiverse Theory is essentially a theory that predicts the existence of multiple universes each tied together and all of them different in some way. This could be what we find at the edge of our universe and you just might be able to say hi to you two, but that's quite unlikely. Physicists still, unfortunately, have not been able to prove this as factual.

The thought of infinite Universes begs so many unanswerable questions: what are in these Universes? How are they different from ours? What is Earth like in all of these Universes? Perhaps we may find a way to enter these other Universes, but even then how would we know where we were? Too many questions.

The Big Bang Theory

The Big Bang Theory states that there was a massive explosion that released debris, particles, etc. outward and began the indefinite expansion of the Universe. However, this is a theory, it's just widely accepted. If you ask most people how they believe the Universe was created they will either say The Big Bang or God. I don't want to get religious but in all actuality, we're practically clueless. The only question I have is: who/what caused the explosion? And this brings the thought that a God or some celestial being may have gone haywire on their science experiment and accidentally created a Universe. A common accident.

The Information Paradox

T he Information Paradox states that information acts like energy, it cannot be created nor destroyed. However, when a black hole reaches its final days, in billions and billions of years due to Hawking Radiation, it will theoretically disappear leaving nothing but darkness in its place. So where does all of this information go? This is another question that plagues the minds of astrophysicists and if Stephen Hawking couldn't answer it then we're screwed.

I suppose it would make sense for the information to be launched outward into the depths of space with no practical pattern, but what would the information look like and what would an event of this calliber mean for the Universe itself?

The Uncertainty Principle

N ow, the uncertainty principle deals with a field of physics called quantum mechanics; pretty much, the easiest thing to understand in quantum mechanics is how to spell it, and even that can be challenging at times. The uncertainty principle was produced by Werner Heisenberg; to understand this principle we must first understand entropy. Entropy is essentially the unpredictability of a given system. For instance an electron in orbit around the nucleus of an atom contains entropy. This means the path of that given electron is uncertain, it is random and there is no way to truly know where it is. The uncertainty principle states that you cannot know for certain both the position and the velocity of that electron. The more you know of one the less you know of the other, since if you know the velocity then the velocity would have changed the position and if you know the position then well the velocity has changed. It makes a great deal of no sense but that pretty much sums up quantum physics. But just like order and chaos, the disorder of the electron allows for the order of the system.

Time-travel

N ow, in physics nearly everything in physics yields an absurd amount of confusion from anyone who tries to understand it. And like quantum physics, time-travel is no exception. Being a staple among sci-fi for a fair amount of time, time-travel has taken many forms from a car that can rip through space time from Back to the Future, or a man who can run much faster than the DeLorean. Both of these express similar and equally comprehensive means of time-travel. Quite simply, they both increase speed to a velocity high enough to actually rip through space-time; although this brings the issue of not being able to pinpoint where you would actually land on the space-time continuum since time is constantly moving it would be extremely difficult to land at the desired time. However, we do not yet fully understand the concept of time to genuinely fabricate a way to travel through it. Einstein did discover a relationship between space and time in which the universe can be thought of almost like a plane where space-time exists as one, containing multiple dimensions which would allow both the travel through space and time at the same instance. The only means of time-travel that I could think to be a reality would be this special relativity; since if someone lived in orbit around a black hole, due to the curvature in space-time created by this black hole (general relativity), the person orbiting around it would be traveling through space-time at a much higher velocity as compared to someone who lived on Earth. This difference could mean that 1 year for the person orbiting the black hole could be 100 years for the people back on Earth, resulting in a time-travel of sorts. Soon maybe we all can be Marty McFly and travel through with an old man scientist.

Special Relativity

Special Relativity is essentially the difference in time given a specific point in space-time. For instance, a person traveling through space on the ISS will travel through time faster relative to a person on Earth. This phenomenon is observable only when there are two clocks that can be compared. This also occurs near black holes; theoretically if you stayed near a black hole, given you weren't sucked in, and came back to Earth after a year, Earth would be much different from when you left since to you it seemed like one year but to them it could have been 100 years. A difference of time based on gravitational attraction. Planets are more attracted to you than you think.

Life on Mars

The Red Planet has been a staple in sci-fi for decades. With the Martians, countless movies based on Mars, even the Looney Toons version on the Martian Manhunter, all share one great possibility: life on Mars. Is it so crazy to ponder the idea of such a concept? Life on a planet other than the rock we thrive on? It seems impossible for it to be impossible. How could our ever-expanding universe with constant births of new planets and solar systems fail to create a planet similar to ours? Well, there's a better chance that life is thriving somewhere in the universe. Creating Life is like following a recipe: take some carbon (plus some more life elements), water, a bit of oxygen, heat, and zap it all up with the perfect amount of electricity and you have a life. But there's one thing missing from this list, and that thing is chance; this recipe has an incredibly small likelihood of creating the dish we know as life, the universe got lucky and probably studied culinary arts. If every planet has a chance to follow this recipe, then how could life not thrive on another celestial body. Lucky for us, Mars has a few of those ingredients, and maybe–just maybe, the universe will beat the odds twice and complete the dish.

String Theory

P roposed by a man named Werner Heisenberg, the scientist that inspired Walter White, String Theory is a complex yet simple theory that describes the Universe in a smaller means than the subatomic level. Below atoms and below all of the many many many many many many quarks, there are strings; strings that vibrate and interact with one-another on a one dimensional scale.

This theory is still in its infancy at the time of this writing and thus the tests for it, don't necessarily exist.
But perhaps the world does run on strings; maybe our world is an orchestra and the Universe is the conductor while each planet represents a different song and the meteors and asteroids are notes coming in and out of play.

Chapter 5

A Bit of Philosophy

<u>Philosophy</u>

P hilosophy has been around since the beginning of time, it only came to be when the first organism capable of asking questions asked: "why?". Thus beginning a new era of living beings that would evolve into us today. Asking simple yet complex questions is the very reason we have become the ignorantly omnipotent species on our planet. We ask and, due to human nature, we must have an answer. This led to a whole new understanding of everything we see around us in every aspect: science, religion, etc.. Without the questions asked by our ancestors, who's to say where time would have taken us. We could have evolved into just taller useless humans that lived in caves and grunted. But, as we see, evolution brought us communication, the ability to gain allies and gain for a common purpose. A basis for all human life.

Questions were asked, and to each question must be an answer; philosophy has declined in popularity in modern times—at least from the ways the Greeks pursued it—but that does not hide its importance, for asking questions is what brought us to this point and even if the answers we fabricate are wrong then, well, more questions are soon to arrive, and soon to come: more answers.

Life persists in many ways, and so do the patterns we perceive.

Philosophy will remain one of many patterns.

Reality

W hat is, what is not; what will be and what has been. All a part of our reality. Reality is a concept of what we perceive as real. Our realities could be the same but the way we define them might be different. Stephen Hawking proposed the multiverse theory, which expressed the idea of various different universes among our own; each maintaining its own version of reality. This idea is expressed in shows like Rick and Morty* or movies like the Avengers†, where time can be manipulated to allow a direct pathway into an alternate reality. Reality can be thought of as a constant, it continues as a collective entity of space and time that cannot be altered. If you think time-travel would alter reality, think about how that very event could have already been preconceived by a celestial power. All of this is hypothetical, reality is only as real as we can define it, to a deaf person reality is based on sight, to a blind person reality is based on sound, and to people with both vision and hearing then reality is based on the combination of the two. Our world can be defined, but only if we fully understand it; which means our reality has much more to offer.

* A lovely show on Adultswim.

† A force of heros created by the great minds at Marvel.

Infinite or Finite

A common question throughout the history of astronomy is whether or not our Universe is continuous or if it has an end. The Big Bang states that the Universe has been expanding, and continues to expand instant ever since The Big Bang occurred. However, no one can prove this yet. We could be just a set of Universes sequentially placed in an order; or we could be alone; alone, with a common goal of achieving Universal omnipotence. But what if the connection of Universes were finite? And what would we call a set of Universes to create one massive: Multiverse? Polyverse sounds pretty cool too to be fair, but Stephen Hawking already gave us Multiverse.

Perspective

Our universe is a very large place, and to truly understand how large the universe is we must try and put its vastness into perspective. Earth has a surface area of 196.9 million square miles; one mile is about the length between two street lights. With that being said, 1,300 Earths can fit within the volume of Jupiter, which has a surface area of 23.71 billion square miles. However, 1,000 Jupiters can fit within the volume of our very own Sun, which has a surface area of about 2.34 trillion square miles. But even the most gargantuan star in the Universe has a radius 1,500 times that of our Sun. Compared to the Universe itself, which is 28.5 gigaparsecs in length, or about 93 billion light years in diameter. It's crazy to think just how tiny we really are, considering I am 6 feet tall but our planet is about 8,000 miles tall. Yet the impact we can have on this planet to create such large and influential effects is dumbfounding.

Life

I f we start at the beginning: why are we here? To suppose that God envisioned a universe with such uniformity to which life could produce is magnificently fascinating. What if God designed a Universe with laws that never could support life? If the Big Bang was never brought into play; never beginning the Universe. Then what?

The Universe only exists because we are here to observe it. Similar to Schrödinger's cat* (which is both dead and alive until you open the box), the Universe could both exist or not exist depending on whether there are observers or not.

Life is similar to a book. Every person has their own story, written by none other than themselves. Each day transcribed onto a new page. Filled with good times and bad; eventually comes a change, a change significant enough to begin a new chapter. With new chapters comes new pages, new good and bad events but all in the same continuous story. As time goes by, just as life has its determinants and its hardships, so does the story. But just like the stories of The Flash† or Doctor Strange†, the conflict is resolved and life continues. Some stories will win awards, and others will be forgotten to collect dust with the rest. It all depends on the author. After the story comes to an end and the final page is written, just the Lorien Legacies§, books may have sequels; and the story written by the author has laid a type of groundwork that can be built upon by authors yet to come. These new authors, winning new awards, collecting new dust, all in repetition; fabricated by the story that began it all. The story of the Universe.

* Schrödinger's cat is a thought experiment that illustrates an apparent paradox of quantum superposition.

† Heroes of wonderful stories written in the past from Marvel and DC.

§ Another wonderful book series by James Frey and Jobie Hughes.

Balance

Everything in life must pertain to a balance. If that balance is broken, then whatever it was holding will break with it. This balance in life functions with order and chaos; without chaos there would be no order, and without order there would be no chaos. Take a dying star, for instance, it was once born into the Universe as a young entity that burned for billions of years; until it finally burned out. Burned out all of the order it maintained only to turn into chaos with a massive explosion spewing space debris and elements into every direction. However, this debris, and this new nebula that is now formed, creates a basis for the new life, the new order that is to take on the future of chaos and order, to maintain this universal balance. Without the chaos of the star, there would be none of the order necessary to create a new one.

Imagination

nything is possible within our imagination. It almost seems as though the world we live in was spawned by a very creative toddler who really likes circles.

Our imagination is a stream of neural networks that transfer electricity from one neuron* to another. On the surface at least. If you're like Peter Pan then your imagination can be anything you wish. Neverland could be your home if you so desire. Or your home could be a box on top of Mount Kilimanjaro, whatever floats your boat.

Imagination is powerful; it allows us to formulate our dreams and act upon them as we wish. It provides a ground for us to build upon whether it be for music, science, or culinary arts, it all sparks with an idea from a little place in our imagination. Willy Wonka's favorite place, and the home of the massive chocolate factory. Dreams can come true so long as you build upon that ground you laid when you first dreamt that dream. You could be Peter Pan, or, well, not Peter Pan.

---

* A grayish or reddish granular cell that is the fundamental functional unit of nervous tissue transmitting and receiving nerve impulses and having cytoplasmic processes which are highly differentiated frequently as multiple dendrites or usually as solitary axons which conduct impulses to and away from the cell body

Order In Creation

O ur world is filled with order, especially society with things being as they must and rarely changing in order for things to "function." As creation occurs it typically calls for both order and chaos as a result. With the creation of technology, for instance, came a great world of communication and advancement but brought with it isolation and a new level of loneliness. Think of a world without creativity; think of the boredom and order it would bring. Each person doing the same things over and over until another comes along to fill their spot; sound familiar? Without creativity, order would not have chaos, and our world would be filled with nothing but boredom and arbitrary order.

Our world seems bored, and yes it can be, but think of the many things to do; if you are sitting at home reading this with your mind blank and nothing to do in the world then try something. Draw a massive eye onto a canvas, cook a meal you've never eaten, write a book, film a movie, create something new or recreate something old. With new experiences comes ideas for even more, and the more you do the more you can create, hold on to that imagination that you had as a kid; you only have one imagination after all.

Luck

L uck is something we think to blame when either an extremely unfortunate event occurs or an extremely fortunate one. Luck is seemingly unquantifiable as it is just a means of describing the act of achieving a highly improbable occurrence. Just like changing your headlight then not a week later your check engine light comes on for a completely different issue. But whether luck exists as its own moral force, so-to-speak, is beyond me; I tend to blame things on luck all the time yet I still question whether some days I just run out of it or if I'm just making incredibly stupid choices (which is the most likely case).

Do you think, however, that luck exists? Do you think it was lucky for our Universe to have been created or for you reading this now to have been born? Or is luck really just a way to describe probability. With all of these being said, I would advise against buying that lottery ticket to win the mega-million, as for that you'd have to be, well, extremely lucky.

Desolation

Desolation is something that plagues us all one way or another. It is a hard thing to talk about but an existing topic nonetheless. With a Universe this vast and a planet this minute in comparison, it only makes sense for many of us to feel in such a way. You could sit and think about the extent to this alienation, although I would recommend you didn't but do as you wish.

Imagine an alien on Mars, traveling and exploring the Martian terrain all by his lonesome; do you think he feels the same desolation? I'd say so. It's crazy how we are surrounded by people and yet we feel the same way as the Martian traveler. Perhaps there is something that must be experienced that will rid people of desolation. Only time will tell I suppose. But on the contrary life can bring happiness when around the right people, and if those people are reading this, I appreciate you.

Hopefully someday that Martian will find a friend, maybe even another rover like curiosity could be his pet. Hopefully NASA would send him a spacebud. That's a mission I would support.

Dreams

There are two different types of dreams: the dreams that contain just random arbitrary things with a small amount of things that mean something to you, and there are the hopes and goals you have. We don't really understand dreams; Freud*tried but we all know what conclusion he came to, and scientists still can't really pinpoint why we have dreams. Could it be your brain telling you the future? Or something that happened in the past? Or is it just organizing a great big jumble of thoughts? All plausible ideas but all equally unknown. Perhaps the dreams you have at night highlight parts of your life dreams; they could contain underlying tips on achieving those dreams, or maybe they really are just you falling off a Burger King into a pool of Rice Krispies.

Hopefully one day a scientist will develop a means of analyzing dreams, or maybe, hopefully, not. Our thoughts are really the only thing we have private in this world, and dreams we, more or less, can't control.

Dreams are important, not just the arbitrary ones but the ones you pondered up when you were little; they're easy to forget but they keep you connected to the purest form of happiness you might ever have; some childhoods aren't the best but before you grasp the idea of the world and when you can still play video games all day without a worry: that type of happiness. So, I keep those dreams that I had and still have, and who knows—maybe someday they'll come true.

* Sigmund 1856–1939 Austrian neurologist and founder of psychoanalysis

Infinite Universe

I magine becoming an astronaut in the year 5024, the technology is incredibly advanced and space-travel is practically child's-play. Your society inhabits many planets in which you visit each day for training. Once you are ready, they send you on and a team on a mission to discover the "barrier of the Universe." You begin your journey, traveling worm-holes that your ship creates that would achieve faster-than-light travel. You and your team continue this journey but after a few centuries, you come to find that the Universe doesn't seem to end; you only discover more planets, stars, galaxies, etc.. The Universe is seemingly infinite.

What if the Universe were infinite? What would happen if the Universe expanded to a size of infinity? There would probably be more aliens than we could count, and we've yet to find any other than our own. This would answer many questions among astrophysics, but spawn a prodigious amount of new ones.

Dreamworld

I magine a world; one of your own creation. Where every-thing was just how you wanted. Every item was in place, you lived in your dream way with a dream life. What would it be like? To live in a world with no hardships, no bad things, just pure joy. Would you be able to experience that joy to its full potential?

Now imagine a world with everything that you despise; nothing went the way you wanted and everything was just something you hated. Would you ever be able to be happy?

The thing about these thoughts is that, if you lived in your dream-world you would eventually become sad; the world would go from joyous to bitter and stale. While if you lived in the oppos-ite, you would eventually become happy; the world started off as stale and bitter, so the only way to go is toward joyous. Just keep moving up.

Chapter 6

Some Randomness

Infinity Stones

Considering I've been reading Peter Pan I don't really have much related to science or physics.

Anyways, there are 5 infinity stones within the Marvel Universe: The soul stone (orange), time stone (green), space stone (blue), mind stone (yellow), reality stone (red), and power stone (purple). Each stone has their own specific power, I would explain each power in detail but that would take much too long.

The concept of the stones is quite intriguing, but it does confuse me how exactly they work; I suppose they hold a certain amount of directed power and only a specific device that harness that power would allow them to be active. There is much to learn in the Marvel Universe but oh well. But I still would love to know how exactly the stones came to be; perhaps I need to read into the lore a little more than I have already.

A New World

W hat would a doppelgänger of Earth be like? Nasa has discovered many planets similar to that of Earth that are within the "goldilocks-zone*" of their home solar system. This means that liquid water can form, as well as a sustainable atmosphere and potentially life. As one ponders the idea of a new world it seems crazy to think that the sunrise might be a different color, there might even be two suns like Tatooine from Star Wars, or instead of one moon there may be over 50 moons to gaze at in the night sky. The atmosphere may be thriving with mainly oxygen rather than nitrogen and there could be species that have yet to even be imagined by the most creative toddlers. A new world would be incredible; to think that a new world would not have the Grand Canyon or Mount Everest, but instead new wonders of the world for people to explore and discover. A world filled with vast clean oceans and the air might even smell different. Even the gravity may allow us to jump higher, and some of us might finally be able to play basketball.

Only time will tell whether or not we do travel to one of these brothers or sisters of Earth, and what a time it will be for the ones who make the journey. Lets just hope Superman doesn't get there before us after Krypton is destroyed.

---

* Lying in or being an area of planetary orbit in which temperatures are neither too hot nor too cold to support life.

† Superman's home planet, discovered by DC.

Minecraft

I'm not sure if I'm allowed to write about this in this book but I still had to nonetheless.

Minecraft blew up all over the world for its simplicity and seemingly unlimited creativity. Whatever your mind thinks of, Minecraft has a way for you to build it, and it's up to you to find that out.

Minecraft allows us to express your creativity without actually having to go outside. We can build a house, raise cattle, and create a farm in a matter of hours. The beauty of this game is unspoken as it gives so much joy to millions and yet it is just a world filled with blocks, but your mind allows you to make it so much more. Keep crafting, maybe they'll make a mod so you can play in real life.

Magic

In our world, as far as we know, magic is not but an illusion or a trick that makes us believe there is another force at play. But in the world of sci-fi it is anything you could imagine it to be.

In the world of Harry Potter it is the casting of spells and brewing of potions; while in Doctor Strange it is a learned discipline from a rather frightening bald woman.

Imagine a world with magic; what would it be? Would you have magic crystals that give a person whatever they desire, or would you put magic in the form of ancient staffs created by our ancestors that control certain elements. The possibilities are seemingly endless. In our world magic is almost everywhere; anything we don't understand we seem to blame on magic, for instance how the remote gets from the armrest of the couch to in-between the cushions must be the doing of the White Witch from Narnia. But as our time continues maybe someday we will discover our own infinity stones, or develop our own magic wands. One can only dream at least.

Tea

Tea is a very great something, there are so many different kinds that all cater to different needs or that just taste good. It's much better for you than coffee and I'm much happier to have a tea addiction then, well, a coffee one. If you are stressed a nice cup of tea and a good book is a remedy that will work 99.9% of the time. You really can't go wrong with tea; unless it's tiger spice or something then it's just horrendous, but other than that.

And to think that the tea we drink was once spawned from a plant that could've spawned from another plant millions of years ago. Much before the eastern countries began to harvest this as a commodity. (Pulled a sneaky on ya). The same plants that developed from materials left on our planet from space dust that fuzed together in order to form our lovely planet. Something so natural and pure that we get to enjoy. As time goes I'd love to grow the leaves myself and make my own homemade tea but only time will tell. Just as time will tell whether or not we continue to drink tea. For all we know, in the near future, people could be using caffeine shots or something of the sort to keep them awake and take them away from drinking coffee or tea. And who's to say that in the future we still have things like tea? Perhaps the world around us will grow so dull and unnatural that we create our own virtual realities that only simulate the experiences we have each and every day.

With that in mind, have a cup of tea, or coffee, whichever you prefer. And remember to enjoy the little things in life, while you do still have them.

Confusion

"Inability to think or reason in a focused, clear manner."

If you search for 'confusion' on Google it comes up as an illness, thanks Google. Everyone has been confused some time or another. It plagues us all to the point where we just forget how to function. Confusion is normal and it often leads to discovery since if someone is confused by a given topic or idea then they tend to expand upon it in order for them to understand it. Unfortunately not everyone can do this and we end up confusing ourselves even more to the point to where we wish we could break our pen but we don't since it's a nice pen. Nonetheless the universe is a confusing place, but that is not to say that nobody can understand it; if you want to understand something, you can. Our brains are extremely powerful; the most complex systems in existence sit right on top of our shoulders. It is quite incredible. And yet this system still can't solve 13+54 after finding $y = mx + b$.

Anyways.

Many people will always be confused and the rest will be slightly less confused, but in the end, if we persist, then we might understand. Unless you wish to understand relativity then I wish you good-luck.

Colors

E verything has a color. Well, except black holes, but that's an entirely different topic. Objects that we see absorb all forms of light except one or maybe a mixture of a few, that reflection is the color that we see. If an object is black then it absorbs every color and reflects none, if it is white it absorbs no colors and reflects them all.

Why do things have colors? Everywhere we look is some sort of color, sorry if you're color blind, seeing in grey is cool too. But our world has had color since evolution brought us cones. Cones are receptors in our eyes that interpret different types of colors once light enters our eyes. These cones are only limited to visible light, as are the rods in our eyes, that deal with seeing in low-light. We do, however, have an absence of many cones in our eyes; for instance, since we have a shortage of violet receptors, we see the sky as blue as opposed to violet. Many other species would see the sky many different colors than we do, but we only know it as blue. What would life be like without color, for every living being? What would it affect? I guess it would suck seeing a grey rainbow after it rained.

Cold

As we go into the winter months, except for Australia, jackets come out and the world is just cold, at least for us. The atmosphere gets colder and colder as our position in space changes, we get further away from the Sun as we make our revolution and the Sun, therefore, provides less heat. Slightly less at least. Water in our atmosphere builds up and changes the temperature of the air, as do large bodies of water. The water in our atmosphere becomes rain and, if cold enough, even becomes snow or hail.

Our planet, however, is nothing close to the cold of space. We have an atmosphere, while empty space does not. The temperature of space is roughly 3K (Kelvin), and is decreasing indefinitely as the entities in the Universe spread apart. With that being said, what if the Universe reaches 0K (Kelvin)? Would particles stop? Would the entropy of our Universe stop increasing? Would there be anything anymore? It is strange but intriguing, maybe someday matter will reach absolute 0, and perhaps that is the entrance of matter to the world of dark matter. Just remember, winter is coming.

Food

Food is possibly one of the best things Earth has to offer; no matter who you are, there's a food that you love to have. The possibilities of it are just endless and the amount of combinations is just amazing.

In perspective of the Universe, what does the Universe consume as "food"? Is the Universe a living and thriving being? I'd like to think so. And I feel as though the Universe would like to snack on something like the dreams we have or something of the sort. Since, the Universe is somewhat a figment of our imagination as we have never, nor will we be able to view the Universe in its entirety; unfortunately, we are only on a Planet and can map it out, again, using our imagination and the dreams we have just added to the monstrosity that is the Universe. Well, dream something good for the Universe to eat up tonight.

Dimensions

A llow me to break the fourth wall: you are currently reading this on a piece of paper. The funny thing is, I am currently writing this on my laptop hoping one day that you are reading this on a piece of paper inside a physical book. It is strange to think that the energy I am currently using is being held in place with an idea from my mind inside of a book that you are still reading. The energy, trapped forever, sits on this page waiting–hoping one day to be read and comprehended. And that, is exactly what you are doing right now; and I thank you for that reader.

Please turn the page before we both go insane.

Why

The question 'why' is a common one to ask. The answer may not be what we expect or it might be exactly what we expect. But this question is essentially the basis for science and all its discoveries since experiments are all based on that question: "why?".

So why do things happen the way they do? It depends on the subject but each thing will have an explanation, and our hunger for knowledge is the perfect plate for the question to be served on. But there are many questions we can't answer, and many of our answers lay incorrect; answers that have hidden among society as correct but have yet to be proven wrong. The next time you look around at the world or notice something, ask why and see where your mind takes you.

Chapter 7

A Hint of Mathematics and Technology

Probability

W hat are the odds that you, me, that dog you always see in your neighborhood, had to overcome to actually be here. Think about the odds that Earth was formed, at just the right temperature, with just the right ingredients at just the right time with a planet like Jupiter to block it from space debris or a Sun like our own not too hot or not too cold. All absolutely perfect. And not to mention the odds of you existing in this current time period; for every single one of your ancestors to do everything one specific, exact way for your parents to one day meet and for them to reproduce at the exact time in which one, just one of the trillions of sperm cells happened to get by and fertilize the egg that is you. Everything might be random, or everything might have a fixed outcome, but to whomever is calling the shots and creating the new entities in the universe, I must say, I'm quite clueless as to how they got everything just right, or wrong? Who knows.

Schwarzschild Radius

T he Schwarzschild Radius was proposed by Karl Schwarzschild and is a radius in which any given mass must con-
dense to in order to become a black hole. For instance,
Earth has a Schwarzschild Radius of about 9mm (0.35in), that's
smaller than your finger tip.

Black holes are essentially gargantuan bodies of mass that collapse on their own gravity and thus confine nearly all of their
mass in such a small space. And those are the small black holes at
their beginning; after millions of years Black Holes devour celestial bodies and gain more mass than anyone could ever fathom.

The equation for this is $R_s = 2GM/c^2$. So if you want to find out how
small you'd have to make your annoying friend, then by all means
plug in the values. You may be able visit the Quantum-Realm.

---

* For instance, my radius would be: $2(6.67\times10^{-11}\text{kg}^{-2}\text{m}^2)(72.57\text{kg})/(299792458^2\text{m/s}) = 1.077\times10^{-25}$ meters!

Robots

Robots are quite an interesting topic. They come in various different forms; from computer AIs to your oven. We see them each and everyday but do we ever notice their capabilities?

Robots have limitless possibilities since they all run on energy and code; code can be essentially infinite and, well, energy is quite abundant in the Universe. With that being said what could we create? Well we could make a giant pizza oven the size of a planet, but I feel only Italy might fund that. We could also make a robot that cleans water—its sole job is to take water and release it again as clean and usable, but many issues might come from that.

What if aliens beat us to it? What if they had a robot that we couldn't even see; a sort of nanobot that recorded our everyday lives. What are the limits for them?

I quite like the pizza one.

AI

A I, or artificial intelligence, is something that most people don't particularly notice. But as we grow in our age of technology, AI becomes more and more prevalent in everyday society. For instance we've seen self-driving cars, bots that can never lose certain games, and language bots that can hold a conversation. AI has grown over time and it has come quite far, but how far will it go? Imagine a time where the world is controlled by AI; your breakfast is already made by a robot that remembers your meals, your clothes were picked out for you by your closet and your groceries have already been ordered and delivered. The life of laziness I suppose.

Imagine a society in which AI is the only functional piece, while the humans sat back and used the society as entertainment but leeched off of it as they needed. What if the humans went too far? While yes computers can be turned off and destroyed, AI can most likely become intelligent enough to reprogram the function of its power button or even build more robots. But that is only if you have an idiotic engineer.

AIs are very useful, if you are reading this you likely use one everyday. Our phones have pieces of machine learning that set reminders or alarms for us and manage things that we do in our everyday lives. There's no telling how far AI can go but I'm excited to see where the road heads; after all a robot that could learn how to make perfect pasta would be pretty great.

Simulation

I magine a video game, one that could simulate life. Each day a string of code that portrays what we see, feel, and do. Perhaps one day EA will make enough money from microtransactions to make a game that could simulate real life. And if this were to occur, who's to say that it wouldn't repeat? Once the game is created, someone begins. In that game someone creates the same game and the cycle continues. For all we know we're dlc that allows you to create your own life, and live through it. A digital sanctuary that takes away from the present of the future. Lets just hope EA goes away before then, we definitely don't need a FIFA 30.

* Electronic Arts: a video game company, that makes great games

Mathematics

Anything and everything can be described and represented by numbers. While math is not the most popular of subjects it certainly is one that is most important, in my opinion. No matter where you are in the Universe, math is always the same; it truly is a Universal language. While different life forms may have other symbols than Arabic numerals to represent them, force still equals mass times acceleration. This is the beauty of math and perhaps this is how we may one day communicate with another species, with our shared love for math. Hopefully they find a mathematician first.

Phones

While it was quite annoying to get these taken away in High School, it is dumbfounding how much power we hold in our pockets. In less than 100 years the human race has gone from barely developing the first television, to creating a computer the size of an entire room. This computer could run some fairly basic operations and some pretty impressive mathematics (I highly recommend giving it some research), and yet we now have computers the size of our hands that can compute everything we need and more. It's crazy to think about how quickly computers are advancing and what the next ones will look like.

Satellites

S atellites, while being some of the most expensive, are some of the most interesting pieces of tech humans have ever created. They are typically built for specific jobs like: observing a planet or a star, testing the atmosphere, or just taking wonderful pictures. These robots can take years to make and are typically really light or really heavy.

Satellites provide information that we humans cannot obtain from our senses alone. They allow us to venture the cosmos without the risk of being instantly frozen or being evaporated by the Sun's deathly radiation. Satellites are quite the treat.

Raspberry Pi

I f you aren't familiar with Raspberry Pi's, they are usually made with dough, some raspberry filling, and a nice pattern to top it off.

Just kidding.

Raspberry Pi's are very small computers that can basically run anything you throw at it; a lot of the time it won't run everything but it can run just like a windows computer. It is used for many things like robotics, and mainly home projects.

The possibilities with these computers are seemingly endless since all you need is knowledge for a program and you've unlocked whatever world you can ever imagine.

Perhaps the Universe is running on a Raspberry Pi.

Speakers

One of the most popular things in the world is music. Similar to math it is something that anyone can understand and enjoy.

In the past the only form of audio was anything you heard in nature; every sound, creak, and chirp came from an external source that you could physically see. However, today we have advanced to the point in which all of those sounds can be recorded, saved, and played with sound waves that our ears are able to understand. From the first gramophone, music has been a huge part of human society, and speakers now allow us to listen to whatever we want wherever we want… For the most part.

The Equation of Life

As I have said before, math is and always will be everything. Everything we know, love and care for can be represented by a tangible equation. Love I'm not so sure about but I guess we can ask Tinder.

Life is complicated, perhaps the most complicated things that humans will ever experience. It is such a broad term yet this one four letter word describes every second of our existence, as we are represented by: life.

Perhaps the next super-computer will also tell us the meaning of life is '42'; or perhaps it will give us an equation. An equation that allows us to plug in the values of our everyday lives in order to receive a number. Now what would that number mean? I suppose we may never know, but I would say it could represent the meaning of our lives to us; a whole number that described our appreciation for our own lives.

As humans the main thing we know is that we are alive and we don't really have to think much about it. But everyday we are affecting others who share that same trait with us: the trait of life. And yet it passes over our mind how much impact we have on life itself, as we are living it. We mean more to the world than we may know, and you reading this mean more to others than you may think. Share that with others, it will help you cherish life.

Chapter 8

Elements of Humanity

Language

L anguage is an extremely important part of who we are. From the beginnings of our ancient ancestors, humans have developed ways to communicate; whether it be hieroglyphs, random grunting or, as my high school English professor would say: lectures. As language was developed, the spread of ideas came along with it. And with ideas came along new inventions like the wheel or pizza rolls.

It's quite incredible how many languages there really are; every country, with some exceptions, having their own distinct languages and even countries that have ceased to exist that died along with a language of their own; books and hieroglyphs that appear like cyphers since the language it was written or created in has nearly been forgotten.

Think of the idea of a universal language; what would it be like? If everyone understood everyone. What would happen to culture and tradition? I suppose one day everyone will speak one language; as english grows more common, it feels as though every country has at least someone that speaks English. The land with probably the least culture, with a language that spreads bacteria.

Hopefully everyone will speak Klingon instead.

Connections

The Universe, with respect to String Theory, is a series of connections. Incomprehensibly small strings that hold the basis for which we live. Just as the Universe has connections, humans share a similar relationship with others. We thrive off of connections and having relationships. Whether it's your significant other, best friend, or both You have a connection; something that you cherish and keep close to you. Without connections the Universe would not be whole, but rather a mess of only disorder, but as we know: with order comes disorder. So yes the Universe has increasing entropy, but it still holds on to its connections; for that is what makes the Universe, well, the Universe. Hold on to what you are connected to, especially if you are zip-lining.

Emotion

"Any of the feelings of joy, sorrow, fear, hate, love, etc.".

E motion is a difficult thing to describe. Especially when you don't even understand your own. However, we all have felt different types of emotions and what it's like to be happy, sad, anxious or mad, and what it does to us mentally. Our brain triggers an emotion by releasing a hormone or neuro-transmitter. This tells the brain to tell us what to "feel", mentally of course. Some emotions can be very difficult to get rid of like sadness, anxiety, or even happiness. But it is all worth the wait as happiness and joy come along as a constant healthy feature of the mind. A cycle of sorts, where only happiness can follow sadness and vice versa.

Emotion is strong and is very difficult to control, and is still not entirely understood. But allowing yourself to feel as you need is important; as is a pursuit of mental health and a better control of your emotions. Perhaps aliens have no emotions, or perhaps not even a brain. I suppose only time will tell.

Sleep

S leep is definitely one of my least and most favorite things of all time. It is my least favorite since it takes about 8 hours of my day every day that could be spent living and enjoying life, or at least trying to. But I love it since, well, it's wonderful.

There are three stages of sleep: as you fall asleep, you have REM sleep (Rapid eye movement), light sleep and deep sleep. You constantly flow through these cycles while you sleep and if the cycle is disrupted then you either wake up in the middle of the night, or you wake up in the morning not feeling well rested.

Do you think aliens sleep? I suppose it would make sense as most creatures on Earth sleep, but of course this is all theoretical and nearly anything is possible.

Maybe they sleep while they are awake, who knows?

Sleep almost seems like someone logging off of a video game; the player plays for a day then disconnects until he logs back on the next day. Perhaps that theory should be saved for a future post. Sleep is just our way to recharge our battery, heal and grow not just physically but mentally. If you sleep on an idea, a new thought may come to you in the morning. Maybe one day people will never have to sleep; when the world is filled with "Visigoths*" who only wish to work and gain power and possessions, and sleep only cuts into their time of work. Hopefully that day will never come to be, life is to be enjoyed and fulfilled—if that fulfillment is to work, then by all means: to each their own.

*A civilization known for their barbaric habits and lifestyles

Sleep II

S leep is odd. Your brain shuts off your body and most parts of our memory to recuperate for the next awakening. But what if we slept for entire days or years at a time? We really don't consider or notice how much we sleep, we just do it because, well, it's pretty darn great. Most of our lives we spend asleep, and the rest is just what we remember.

Sleep for however long you need to, and I recommend naps they're wonderful. But spend the rest of that time experiencing, I can hear the birds chirping outside as I'm writing this, and it shows there's a world out there to explore, so explore it. Take those memories that you get and hold them, for they are what make up your life.

Happiness

Being happy is part of being human. Typically, it's one of the best parts. Being sad is difficult at times and it hinders our ability to be happy; in the end we all pretty much wish to be, well, happy. So what makes you happy? What brings you joy? What makes you excited just thinking about it? Take that thing, and pursue it, it will make your life: happier. Bring joy to those around you and allow them to bring joy to you, the world needs happiness but stick with the world in the times where it needs sadness. It's like a cycle, the precipitation cycle; whenever it rains, soon after that rain that brings gloomy clouds and grey skies, comes a bright rainbow and next a nice sunny day. Following sadness comes happiness, and both sadness and happiness are a given in any circumstance.

Adventure

T he Universe is vast; much too vast for us to explore at our current state of advancement. But what is among our technological capability is our solar system. Part of which is beneath our feet as we speak. We don't know everything about our solar system and the planets that compose it, and Earth is no exception. With roughly ~65% of Earth yet to be explored there is still so much room for adventure.

Adventure allows us to understand and enjoy the planet we live on; unless, of course, it's the Mariana trench*: that's just terrifying. But as time goes on we will explore and discover and maybe a dinosaur will jump out from a cave, hopefully it'd be a herbivore. But to be an explorer and take adventures, there is always room for discovery.

*Sea floor trench in the Pacific extending from southeast of Guam to northwest of the Mariana Islands and containing the deepest known depression on Earth, at a depth of 36,198 feet (11,033 meters)

Thought

O ur minds are complex. The most complex thing known to man is sitting on top of our shoulders. We have the ability to think and to ask, to ponder and to wonder. We take ideas and expand and every thought branches into a new thought and it only continues from there.

Our brains mix our thoughts with our emotions. Some thoughts may be strong, some weak, some happy, some sad. But they drive how you think, what you do, and what you say. The emotion it creates can also manifest in your head, taking over your mind completely until your head is filled with only one thought. And the emotion amplifies itself until a resolution is discovered. Our minds are powerful, but what we think is more powerful than we believe.

Creativity

"The use of the imagination or original ideas, especially in the production of an artistic work."

Creativity is everything. Everything we see or use was either fabricated by a Creative Thinker, the imaginative Mother Nature, or the wondrous Universe. Creativity allows us to manipulate our thoughts and ideas into real life things. It can be expressed in music, or across a canvas in a work of art, or through the ingenuity of an invention. Everything is creative, and whether you're creative-minded or not, you have a creative side when it comes to something. Whether it be culinary arts, or bird watching, creativity can be expressed through anything.

As long as the creative strive to create along with the Universe and our wonderful Mother Nature, we can strive as humans to make our lives more vibrant and full of metaphorical colors. Colors that we've never even imagined. And it all comes with that first idea.

Seasons

When you think about a season, you most likely think about your favorite one But it's quite interesting to think of the way in which it occurs; society essentially revolves around certain parts around these seasons. What if we lived on Mars, how drastically would society change?

Imagine Christmas time or Springtime on another planet; what if aliens had holidays? I could just imagine an alien Santa Claus being the leader of an entire planet. But that's just theory of course.

The Earth is very strange, but it's amazing how much it changes when the seasons switch. Plants change, animals move, we get depressed, it's all a cycle. But what if we lose a season? Soon we could lose winter to the warming of our planet. If I were you I'd plant a tree, or at least donate to an organization that can do it for you: the Arbor Day Foundation. Gaze at a tree next time you're outside, you might see something beautiful.

The Heart

L ove is powerful. I describe it as a force, one that we cannot see nor measure. It's a force that pushes and pulls on emotions and attracts beings with the same love.

What is the heart? One big muscle; the great muscle of the human body. It has four sections for pumping blood and can be thought of like the main train station for blood transport. But what's inside the heart? Behind the blood and the tubes and the tissue, what is it? It's almost as if the heart is the container in which our spirit or being is held. Locked within a shell that cannot be opened with ease, metaphorically or physically.

Our hearts hold our souls; they nourish them and mature them until they can experience love. But, who knows, an anatomist could probably tell you more about what's inside, I can only theorize about what's hidden beneath.

Chapter 9

Why?

<u>Why are we here?</u>

A question that has perplexed the minds of many for centuries. It would seem we are here in a game; a game that we cannot escape that binds us to the Universe in which we must develop our own purpose. I suppose if this were true it would be interesting to meet the developers of this game and what exactly their motives were. It may seem that our lives are just ways of providing entertainment to galactic beings; an entertainment that allows us to watch from afar as we destroy our planet and separate ourselves from our own species. An odd question indeed but one that should not go unasked.

Why do we eat?

I suppose the answer is quite self-explanatory, that we eat to fuel ourselves as any form of life needs some sort of energy source in order to continue. But it's interesting to think that life was developed in such a way that caused carnivores to eat other living beings to sustain their own life. It's as if the Universe put beings into a spherical cage and whoever won got to live. Perhaps there is another way of gaining energy, a way that would not require consumption or anything of the sort. But rather a source that allows us to feed off of thoughts and wonders. Going to restaurants would be people sharing ideas and it would seem they grow more full the more ideas they share, as it must have some sort of limit. This is a society I feel would be greatly advanced, and perhaps one exists already in the multiverse.

Why do we persist?

T his is not something that applies to all, but it applies to most. If you are reading this right now, then you, like me, have a hunger for knowledge and are persistent in your pursuits for learning about all that you can. Staying persistent is in human nature, it is a trait that grows alongside confidence; something that is usually hiding beneath the surface of someone's personality. Anybody can attain this trait; it's just a matter of comfort and again: persistence. It allows us to build upon ourselves and those around us by not letting ourselves give up, and while there are times where we should take a break or 'give up,' it is in our nature to continue forward to achieving our goal. This is what defines the curious; the ones who seek and conquer the world around us, and to you that fight alongside your peers, and yourselves, I commend you; take your curiosity and allow it to thrive in this beautiful world, as it will allow you to persist and achieve that in which you dream of.

Why do we feel?

E very person feels emotion; some stronger than others and many seem to stay out of touch with that of which they feel. But overall, why do we feel any of this at all? It would seem that emotions are a biological way for your brain or your body to tell you how you think about something. For instance, if you had a date coming up you feel excited or if they didn't show up you feel sad. However, that sadness may linger for a while and that still gives power to the puzzling question asked above.

Feelings are strong at times and while it may seem they have power over us, after all, we are the ones giving them the power. It is crazy to think that everything we have ever felt has come from chemicals inside our brains, along with a physical event that catalyzed a reaction among those chemicals.

Perhaps we feel, as it helps us find our purpose.

Why do we hate?

The phrase "hate is such a strong word," has so much truth to it; when people say they hate you or if you say you hate somebody it is typically because they have done something extremely wrong in your eyes. But why must humans hate one another in such a strong way? To hate other people would seem to be something unintended by the Universe. With respect to life, we all live on the same planet in the same galaxy in the same galaxy cluster, and yet we cannot get over our differences and appreciate each other's existence. It is understandable if someone did something extremely wrong as stated before, but does that constitute hatred? Hatred is what starts wars and battles that, from above, look like play among children that cannot get their way. A world we live, that I may never understand, but I hope no one has caused you to hate them.

Why do we seek joy?

For mostly all it would seem that people enjoy feeling good. Or, in other words, people enjoy when their brain releases dopamine and allows them to feel that feeling of happiness inside their body. It is strange, however, as these feelings that we feel consist of chemicals and catalysts with a few instructions; our brain is the chemist and the emotions we feel are just the solutions that result from the experiments.

It would almost seem as though the brain reacts with the heart in which the brain creates a reaction and the heart either increases in bpm or it grows sad and can either decrease in bpm or increase if scared. Other parts of the body do the same as if to predict for what is to come next. I suppose joy is that of what our heart deems worthy of bpm* increase, while our minds are not making our eyes flow with water.

Only theories.

* Beats per minute

Why do things hurt?

Pain is something that is seemingly unavoidable by people; it would seem the Universe has put pain in to play to counteract the effects of joy or happiness to keep the balance of the Universe. However terrible it may be, it is an essential part of our existence. For a more philosophical view, feeling pain would allow us to properly understand and appreciate the feeling of joy. Being able to appreciate that happiness is something I feel all should strive for; while yes pain is horrendous and can be detrimental at times, mental pain is a sign of growth and can allow for new appreciations to arise in life. With that being said, be careful with your mind. Pain and pleasure are not things to play around with as they are way more potent than it may seem. Be safe in the world of hurting joy and learn from your experiences.

Why are celestial bodies spherical?

W ithin the basic laws of physics, it is known that in space bodies of mass attract other bodies of mass. Looking at planets, stars, and other celestial bodies, we see that these are all objects of great mass all located in one spot in the Universe. Each of these masses elicits a gravitational pull which their mass also has to abide by; thus all of the particles and bits of their physical existence is being pulled inward in all directions. This would allow for one shape to be created once all the mass is pulled in: a sphere.

Perhaps in a parallel Universe gravity is not omnidirectional and instead only allows for objects to be pulled in four directions resulting in shapes like cubes or rectangular prisms. An interesting Universe indeed, but one that would definitely be accepted by the early sailors of the 15th century.

Why do we ask why?

For humans, over time, we have been in pursuit of knowledge and power. Being able to understand the world around us has allowed for the very advancements that we see today. Without asking 'why?' humans would not have been able to understand why we need food or water, why fire exists, or why those people keep cutting you in line Starbucks. Having this knowledge has allowed humans to create new things and ideas based on the ones already made by the Universe. The Universe has given these ideas as tools that beings can use to their disposal to create a world of their own. It seems, however, we ask 'who?' or 'what?' more frequently than 'why?' in today's age, but we still leave room for a bit of curiosity. For that curiosity will be the toolbelt we all use to create new things and ideas of our own someday.

Chapter 10

Elements of Life

The Universe

T he Universe is large; the largest thing we know of at least. We can only see but a small part of the universe which we call the "observable universe" and what lies beyond that is, well, beyond our knowledge. Maybe we have brother and sister Universes that will one day collide and make one giant Universe. Or maybe if we make a large enough particle accelerator the reverse flash will run through it and build us a time machine. Even though the likelihood of that is quite small, it's still fun to ponder.

For all we know, Elon Musk could've been right; the Universe could just be one, relatively, massive simulation with distinct laws and motions that make sense to us because we've lived with them for so long. We could just be a Universal version of No Man's Sky.

Many philosophers and cosmologists have wondered at the question "what is the universe?" And as we inch closer and closer to an answer we only open new doors to new possible solutions. Just like the Uncertainty Principle, we cannot know everything—our feeble minds would collapse on themselves and form a black hole if we had all the knowledge of the universe. I suppose that would be quite interesting but that's besides the point. Maybe someday we'll find an answer that will fulfill our hunger for solutions, but even when that day comes, who's to say that a future Galileo won't emerge and convince everyone that we live in an infinitely dense potato that orbits around one dinner plate. I doubt this would persuade the public but you get the idea.

Music

Our world is music, everywhere you go you hear a rhythm or some sort of beat. Music is strong, it can control your emotions and make you feel things that you never would have thought possible. Music can change you, so long as you're listening.

With string theory in play, the Universe could very well just be one giant song; a song that has continued since the dawn of creation and still plays on to this very moment. The strings that make up this Universe just vibrate at a rhythmic frequency that we are unable to hear. There is something special about music that I can't quite understand yet. But music is something that is engraved in our minds, activated or not. If the world is set on a rhythm then maybe someday we can all learn how our world was created...through a song.

Chronology

E verybody has a clock; no, not a clock that you would put on your wall or on your wrist. But a figurative clock, a universal one. One that tells you the time of your existence, and to think further than that only brings fear so I'll try to stay away, but this time is yours. No one else has this time but you. Now, of course, society and life make this difficult to be true but at its basis, time is yours.

Time is relative; thus time is specific for each and every person. And what you do with that time is whatever you desire. Of course what you do may have consequences or rewards but, again, at its basis, time is yours. We cannot foresee our future, we can only gain knowledge from the past, which makes our time in the present all the time we have; so do with that time what you desire, ask that girl or boy on a date, try out for the team or pursue your dreams. For what you don't do now, you may not be able to do later.

Continuity

Infinity is a great value. Incomprehensibly great. We cannot truly understand how great infinity actually is. It is bigger than any number you could ever imagine plus some—and it continues forever. It has an infinite continuity; in other words, it never ends. What does that mean exactly? Well, who knows. We theorize that the universe is "infinite" but with respect to what? Time, perhaps; we don't know what happened at the beginning of time or what will happen at the end of it, but something will continue. The question this brings is: what will continue? Perhaps time will truly continue, and the Universe will give birth to another existence, or great emptiness filled with, well, stuff. Only time will tell, after all. I guess our children's children$x10^{999}$, might find out, maybe they'll send us a postcard. I can only imagine what they would say, perhaps something like: "Hey there old man," if they still speak english, "we now live in a star and the Universe is a sea turtle." Or—probably more scientific.

Art in Creation

We create things each and every day. Whether it be a sandwich or modern-day Mona Lisa, we are all artists in our own way. Creation is the key to existence since it spawns from ideas or thoughts that are then placed into reality. The Universe had an idea; to make a void and fill it with matter and let that matter create itself. It then created us. Without creation we would be nothing. As time goes on creation becomes destruction just as order becomes chaos. But creation persists just as well as destruction does. The Universe is an artist and we are the strokes of the brush on a canvas that is our planet. One of many works of art.

Illusion

The world we live in is just one big magic trick. It's more-so magic to me since, well, I don't understand it in its entirety and if I did my brain would be so dense that it would become a black hole.

An illusion is described as a sort of distortion of senses; something that messes with how your eyes process things or how you hear something, etc.. When looking out into the Universe, we might see light even though that star that once shone bright doesn't even exist anymore. Or how our eyes receive light upside down and the picture created is then flipped by our brain.

The Universe is a difficult place to understand, mostly everything seemed like magic or an illusion at some point, the witch trials, until people wondered and figured out the reasoning behind why something is the way it is. With that being said, it is extremely likely that we see something every day but simply ignore it because we've yet to understand it; this could be something like dark matter or string theory. Our Universe is vast and confusing, but ponder every now and then; allow your mind to wonder how things are and don't let your mind settle on the idea that things just work because they just do. Everything has a reason whether we notice it or not.

Experience

The past builds and develops who we are in the present and who we will become in the future. We can manipulate and morph this process to our desire.

We experience things every day, but there are experiences that stick; experiences that set themselves in a memory box in our mind that when opened release emotions of happiness and sadness. Nostalgia. It is strong and it lingers in our memory like a leech on a body of emotion.

Our love can fluctuate from emotion to emotion but nostalgia and memories bring us back to the times where life was good; when life was happy. I hope to make more of these memories and look back on them when I'm older so that I can experience a fragment of the joy I experienced at the time. Enjoy life and enjoy the nostalgia.

Stories

One of my favorite things is stories. Good ones, at least. To tell a narrative and drive the emotion and wonder of whoever is paying attention is just beautiful. And that is exactly what we are doing here; our lives persist as a narrative, entertained and written by not only ourselves but the Universe. Each story is important and unique; having different details and specifics. But they're all stories nonetheless. A plot that you get to write, with people by your side to edit your drafts and revise your final product. Stories are beautiful and the world deserves everyone's story, as they all mean something.

Maybe someday the Universe will tell us its story, and maybe we can even rewrite it.

Pictures

What if every time you take a picture, you open a portal to a new dimension, we just don't have the ability to enter it? That dimension is in your mind, caged within your memory and only you have the key to unlock it. Whenever you look at that picture again, the cage opens and it brings you back to that time and everything that had happened. All the emotions flood back and you can describe exactly what happened to even the most minute of details. What if the multiverse theory only wishes to tell us that our Universe, along with many others, are just memories; captured in a frame or in the mind of just a single person.

You cannot forget your memories, they only become lost and hidden behind the newer ones you have made. But every time you take a picture, it makes it easier to retrieve that memory and bring back that emotion. Every time you take a picture, you open up a new Universe in your mind and you store it as a memory; hopefully it's a happy one.

Love

There are some things in life that take your breath away; some things that make your insides smile so much that you can't help but smile on the outside. These are the feelings of love, along with many others of course, or at least what my young feeble mind perceives it to be.

Without this feeling where would humans be today? No true connections, no appreciation for another person, just a bland life lived with oneself and perhaps a few boring exchanges with other people.

As much as the world talks about and presents love, it seems that it is something that is thought about very little, at least in a scientific view. Love, realistically, is just a series of chemical reactions in your brain, along with other places, that causes you to feel a certain way towards a tangible thing. Whatever that thing is is completely subjective to the person but it's felt by all nonetheless. Love is a very important part in who we are as it allows us to express and share an emotion that brings joy, sadness, and allows for learning for the most powerful parts of our emotions. Without love, people would be heartless; the world would be a completely different place.

If you're reading this and have yet to ask out that special person, go for it—if you think the time is right.

Cost

Everything costs something but does anything cost nothing? Cost assigns a value to something and implies that it has a price which is not necessarily monetary. Everything has a cost—everyday you sacrifice a bit of sleep or sanity to show up to school or work, that's a cost. That food you ordered or that video game you bought came at a cost, it's unfortunate but it keeps balance. If nothing had a cost then how would the Universe maintain equilibrium? The Stars in the sky convert Hydrogen into Helium under nuclear fusion, but every time this occurs the star's core gains Iron. The longer this goes on, the more dense the Star's core becomes and eventually it collapses in on itself due to its own gravity. If this didn't occur, then stars would grow indefinitely, disrupting the Universal equilibrium.

So yes, at times cost is obnoxious and can get on our nerves but it is simply a sacrifice for things yet to come, sometimes you'll buy bad things but most of the time you buy good things. Every purchase is just a transaction with the Universe, you're buying the future and the Universe will ensure its equilibrium.

Reality

Our minds look, and work the same, but we think differently. No two people have the exact same brain, but they are able to think the same thoughts. Does that mean they must have the same realities? I'd say no, but to each their own.

What if every person had a different version of reality? What would be different? How would it differ from yours? It all depends on the brain; the snowflake of the biological world. Take time to compare realities with others, you might uncover a piece of life you never knew existed, or your brain just didn't process. Just don't ruin the reality of others, it's the only one they've got.

Wonder

T he world we live in is full of wonder. Everything and any-thing can spark a reaction in our minds and be the cata-lyst to the fabrication of our thoughts and ideas. And as we live and think, we wonder; we wonder how this happens or how that does. No matter what it's for, humans are curious. To wonder is to have an open-mind; a mind that yearns for answers to the questions it asks itself, that you ask yourself.

Curiosity brings reason to life. It makes it so that the world around us isn't just the world around us, but rather it is an entity composed of the elements and ideas we see just by looking out the window or walking on the sidewalk. As we grow curious, as our mind wonders, these things grow colorful, you notice them. Like adding a piece to a puzzle, the world looks more complete. Next time your mind is free, fill it with the world around you, see what you find.

Wonder II

D o you ever wonder? About anything? About everything? A wonderful human trait is that we have the ability to ponder; to think outside our reality and wonder what other realities would be like. It's truly incredible.

When you wander the forest you might wonder what animals are lurking about, or you might wonder if there are any secret caves awaiting your entry so that they can present to you a trial. If only we were all Link.

What do you wonder most about? Imagine what other people wonder about; the realities of others are quite different from your own, or at least, I assume.

Wonder. Take time to ponder and ask questions about things that confuse you, or about anything at all. The world is a beautiful place if you ponder its beauty.

A Conlusion of Sorts

I 've never been all that great with conclusions, so here we are.

I don't want this to be the end of the story, so it won't. Here I am going to deem this the beginning but it won't be the beginning of the book. No. This will be the beginning of your adventure; this will be the beginning of your curiosity and now that you've gone through these chapters and embarked on this wondrous journey with me, you can open your mind to the wonders that surround you each and every day. And if your mind has been open then I commend you.

This life is not an easy one to live; there are many days where you won't want to wake up or days that beat you down until it feels like you have nothing left to get you back up, but you do. You have that mind on those shoulders of yours that is yearning to learn more about the world it exists in. And with that said I encourage you to explore; adventure into your wildest and most amazing dreams until you find and create new ones. You'll meet incredible new people on the way and you'll learn more than you could ever ask for. Like I've stated in parts of this book, nothing exists without balance and expecting a perfect life or a perfect existence is impractical, improbable, and impossible. Even the people that are looked at who seem to have 'perfect' lives deal with things that are terrible to them, but it seems easier than the struggles of the bystanders watching.

Nothing in life will go the way you want it to or the way you expect it to, and this was something that I've only recently began to truly understand. For years nothing has gone exactly the way I planned it and the mind of an OCD high school/college student hated this idea, but I've realized that it isn't going to go away. To except change is to except having an open and free mind and to allow new information and new experiences. The only harm that can be done by change is the harm that you allow.

I can't say that everything I say is truthful or everything I say is meant to be taken and accepted, but it is to be taken and interpreted. I hope that throughout this book the words and ideas that I have expressed to you at least made you think and made you question the world around you, and not in a bad way. I am not always right and this book is a great example of that, I have only expressed ideas that interest me, and thoughts that have manifested in my mind that are meant to be taken any way that seems fit.

Being curious is the one thing I can say I enjoy most about life. Going outside and dreaming about what wonders might be waiting out there for me to explore and discover brings me more motivation than I could ever need, and lately I've lacked that fuel quite a bit. There is so much for everyone to explore and I hate to see people giving up on their dreams just to allow for numbers to increase in their bank account. I understand that this world has been built in such a terrible way that a lot of people hate their life so that they can pay bills and raise children they can't afford, in a broken house they can't afford to fix. But a life of someone who is curious is rich enough to where they would find ways to fix their house, explore new opportunities and provide for their families. I don't have children nor do I have my own home but as an 18 year old writing a book out of passion and curiosity, these are just observations I've made and conclusions of come to.

I hope in the future I can write more for you all and help keep that curious part of your wonderful brains thriving and healthy as it should be. With the good and bad parts of life, there should be one thing that everyone can enjoy to have; one thing that everyone can afford and that all can obtain:

Curiosity

Stay Curious

Acknowledgement

Writing this book has been quite insane; but seeing the final product and being able to begin my journey as a writer is a wonderful thing. I am very excited to continue improving my craft and to hopefully make more books that everyone can enjoy. But above all, I hope you learned something from this book or at least see the world as the beautiful place it is if you open your world of curiosity. I must thank my friends who I shoved paragraphs at for them to read and to bounce ideas off of. As well as anyone who has told me that I have potential or that I can do more because I'm on the path of doing all that I can. So thank you, for making a dream of mine come true and reading my first ever book. And remember:

Stay Curious

All defenitions in this book were provided by merriam-webster dictionary

WWW.MERRIAM-WEBSTER.COM

Printed in Great Britain
by Amazon